THE POETRY OF BERYLLIUM

The Poetry of Beryllium

Walter the Educator™

SKB

Silent King Books a WhichHead Imprint

First Printing, 2023

"Earning a degree in chemistry changed my life!"
- Walter the Educator

dedicated to all the chemistry lovers, like myself, across the world

CONTENTS

WHY I CREATED THIS BOOK?

Creating a poetry book about the chemical element Beryllium offers a unique and creative approach to exploring science and art. Beryllium, with its atomic number 4, possesses fascinating properties that can inspire poetic imagination. Through poetry, I can delve into the elemental nature of Beryllium, its atomic structure, its significance in the universe, and its various applications. This book allows for the fusion of scientific knowledge and artistic expression, bridging the gap between two seemingly distinct realms and encouraging readers to appreciate the beauty of science from a different perspective.

ONE

SCIENCE AND BEAUTY

In the realm of elements, a gem of light,
Beryllium, a treasure shining bright.
Its atomic number, four, reveals its worth,
A metal rare, with secrets to unearth.
Within the Earth's crust, it lies concealed,
A jewel of nature, its beauty revealed.
A silvery-white metal, strong and light,
Beryllium's essence, a marvel in sight.
Its electrons dance with grace and charm,
A symphony of particles, an atomic farm.
With a nucleus small, protons, and neutrons,
Beryllium embodies nature's splendid tunes.
In stars afar, its creation begins,
Born from cosmic fire, where life begins.

Through stellar fusion, elements collide,
Beryllium's birth, a cosmic ride.
Its presence felt in precious gems,
Emeralds and aquamarines, nature's gems.
Beryllium's touch, a vibrant hue,
Adorning the world with colors anew.
With strength unparalleled, it finds its place,
In alloys and materials, a foundation to embrace.
Aiding in aerospace, aeronautics, and more,
Beryllium's might, a scientific lore.
But caution we must exercise, for it holds a risk,
In its powder form, dangers brisk.
Lung diseases lurk, a silent threat,
Handling beryllium, a task to fret.
So let us marvel at this element rare,
Beryllium's elegance, beyond compare.
A gem of the periodic table, a wonder to behold,
In science and beauty, its story unfolds.

TWO

ELEMENT RARE

In the realm of elements, there shines a rare metal,
Beryllium, elegant and precious, a marvel.
Silvery-white in appearance, it gleams with grace,
A jewel in the crown of the periodic space.
Beryllium, a paradox of strength and light,
A feather in the wind, yet a titan in its might.
Its atoms dance with vigor, bonded tight,
Unyielding yet delicate, a captivating sight.
Within its core, secrets of emerald lie,
For Beryllium births these gems, money can't buy.
Emeralds, a symbol of nature's vibrant hue,
A gift from Beryllium, so rare and true.
Aquamarines too, with their tranquil allure,
Beryllium's touch, like waves calm and pure.
In shades of blue, they whisper tales of the sea,
A testament to Beryllium's alchemy.

Beyond the realm of gems, in industries it thrives,
Aerospace, aeronautics, where Beryllium strives.
Lightweight and sturdy, it takes to the skies,
Enabling dreams to soar, reaching new highs.
Yet, in Beryllium's powder, caution must be heeded,
For health risks await, if not properly treated.
With reverence and care, we handle this treasure,
For its wonders are boundless, beyond measure.
Beryllium, a symphony of wonder and grace,
An element that enchants, leaving none in its trace.
From gems to industries, it leaves its mark,
A testament to its power, both light and dark.
So let us celebrate this element rare,
Beryllium, a jewel beyond compare.
In its elegance and wonder, let us find delight,
In the beauty of science, shining clear and bright.

THREE

ELEMENT OF GRACE

In Beryllium's realm, a paradox unfolds,
A dance of secrets, stories yet untold.
A metal both sturdy and featherlight,
Its presence holds allure, a captivating sight.
Within the earth, its atoms reside,
A treasure concealed, hidden deep inside.
Emeralds and aquamarines, gemstones so rare,
Beryllium's touch, their brilliance to bear.
In aerospace's realm, it takes to the sky,
Aiding innovation, soaring up high.
Its strength and lightness, a marvel indeed,
Crafting wings of progress, fulfilling our need.
Yet caution must guide our every step,
For Beryllium's touch, a danger we can't forget.
In its embrace, the unseen peril lies,
A warning to heed, for health's compromise.

A paradox it remains, this element of grace,
A creator of beauty, a guardian in space.
In labs and industries, its power we seek,
But reverence and care, we must always keep.

FOUR

HEALTH RISKS LURK

In the realm of elements, a gem so rare,
Beryllium shines with an ethereal glare.
With atomic number four, it gleams,
A radiant beauty, beyond our dreams.

Found in crystals, pure and bright,
Emeralds and aquamarines, a stunning sight.
With hues of green and blue, they adorn,
Beryllium's presence, like gems reborn.

In aerospace, it plays a vital role,
In lightweight alloys, it takes its toll.
The wings of planes, soaring high,
Beryllium's strength, it does imply.

But caution we must exercise,
For health risks lurk, in disguise.
Exposure to Beryllium, we must beware,
Protecting our lungs with utmost care.

So let us marvel at Beryllium's grace,
Its dual nature, a paradox we embrace.
In gems and industry, it shines so bold,
A treasure to cherish, but handle with hold.

FIVE

WONDERS OF BERYLLIUM

In the heart of Earth's embrace,
A gemstone hidden, full of grace,
Beryllium, a treasure rare,
With elegance beyond compare.
A whisper of palest green,
A secret beauty yet unseen,
In emerald's glow it dances free,
A gemstone's crown, a sight to see.
But Beryllium, more than a gem,
A metal strong, a shining emblem,
In aerospace it takes to flight,
Guiding souls through the darkest night.
With strength and lightness, it finds its way,
In satellites, it never strays,

A silent hero, steadfast and true,
In the vast expanse of endless blue.

But caution, dear friend, heed the call,
For Beryllium, a risk for all,
Its touch, a danger, invisible and sly,
Inhale its dust, and troubles lie.

Protective gloves, a careful hand,
In laboratories, where it may stand,
Respect its power, its potential harm,
For Beryllium, a double-edged charm.

Yet in its essence, we find delight,
A shimmering jewel, shining bright,
Beryllium, a marvel to explore,
A testament to Earth's ancient lore.

So let us celebrate its grand design,
The wonders of Beryllium, so fine,
With reverence and care, we shall embrace,
The beauty of this element's grace.

SIX

TOXIC EMBRACE

In the depths of the Earth, a treasure gleams,
A gemstone of beauty, Beryllium it seems.
With hues of emerald, aquamarine, and gold,
Its presence in crystals, a story untold.

From the heart of mountains, it rises high,
A marvel of nature, reaching for the sky.
In aerospace realms, it soars through the air,
Lightweight and sturdy, beyond compare.

Industries embrace its formidable might,
For Beryllium's strength is an exquisite sight.
From satellites to engines, it lends its grace,
Unyielding and resilient, in every space.

But beware, oh mortal, when you approach,
For Beryllium's touch demands utmost coach.
A cautionary tale, a whispered refrain,
Handle with care, lest you invite the pain.

Its toxic embrace, a silent threat,
To lungs and to life, a perilous bet.
Respect its power, its elegance and might,
But never forget, it demands our foresight.
Oh, Beryllium, a paradox so rare,
A paradox we must handle with care.
In your mystic beauty, we find our delight,
But caution and reverence shall guide us right.

SEVEN

REVERENCE AND CARE

In the realm of aerospace, Beryllium soars,
A metal of wonders, a beauty that roars.
Its strength and its lightness, a marvel to see,
But beware, dear wanderer, for danger may be.
In spacecraft and satellites, it finds its home,
With elegance and grace, it freely will roam.
Its low atomic number, a gift from the stars,
Bestows it with qualities that reach afar.
But heed my words, for caution is key,
For Beryllium's touch can cause harm, you see.
Its dust, when inhaled, can bring illness and pain,
A reminder that beauty and danger oft remain.
Respect this element, with reverence and care,
For its dual nature, both gentle and rare.

Within its allure lies a lesson profound,
To cherish and protect what we have found.

EIGHT

CAUTION MUST SWELL

In a realm of elements, rare and bright,
Beryllium shines with a dazzling light.
A metal so delicate, fragile and pure,
It beckons us closer, its allure.
But heed this warning, dear seeker of grace,
Beryllium's touch can be a perilous embrace.
For in its beauty lies a hidden sting,
A toxic presence that danger does bring.
With a touch, it whispers a cautionary tale,
Of the perils that lie within its veiled trail.
Handle with care, this shimmering gem,
For its touch can inflict a poisonous stem.
Beryllium, a paradox of beauty and dread,
A dance of caution, where respect must be tread.

So marvel at its brilliance, but remember it well,
For in the realm of elements, caution must swell.

NINE

PARADOXICAL ELEMENT

In the earth's crust, I do reside,
A metal with a dual-sided pride.
Beryllium, the name that I bear,
A precious stone, a metal, with flair.
My alloys are strong, my use is vast,
Aerospace and precision, it's no mere contrast.
But beware my beauty, for it comes with a cost,
My tiny ions can have your immunity lost.
Respect me, be cautious, when handling with care,
For my presence can cause berylliosis, a deadly affair.
Yet, in the right hands, I can do wonders untold,
A paradoxical element, with stories to unfold.

TEN

MARVEL AT BERYLLIUM

In the realm of metals, shining bright,
There lies a noble element, Beryllium, so light.
Its atomic number, four, a small delight,
But heed its nature, a cautionary insight.
With strength and lightness, Beryllium shines,
A precious gem, a treasure to find.
But beware its touch, for danger hides,
In its toxic embrace, pain resides.
From emerald's green to aquamarine's hue,
Beryllium's beauty captures the view.
Yet its toxicity, a silent threat,
Inhaled or ingested, it poses regret.
Industrial workers, beware the call,
For Beryllium's touch can cause a fall.

Lungs and skin, vulnerable prey,
Handle with care, don't let it stray.

So marvel at Beryllium's wondrous grace,
But remember the risks it may embrace.
With knowledge and caution, we can prevail,
In the presence of Beryllium, let us never fail.

ELEVEN

SHIMMERING GEM

In the realm of elements, a paradox resides,
A metal both precious and toxic, it hides,
Beryllium, its name, a whispered tale,
With allure and danger, a delicate scale.

A shimmering gem, a crystal so rare,
Its beauty entices, it's beyond compare,
Reflecting the light with an ethereal glow,
A marvel of nature, a spectacle to show.

But beware, dear wanderer, of its toxic might,
For Beryllium's touch can cause quite a fright,
Invisible danger, a venomous sting,
A cautionary note that the winds must bring.

With great strength it binds, in alloys it thrives,
Enhancing our vessels, where progress arrives,
From satellites soaring in boundless skies,
To machines that hum with each enterprise.

Yet, handle with care, this enigmatic guest,
For its toxicity hides beneath its chest,
Respect its power, its presence profound,
Keep vigilance close, let caution surround.
Oh, Beryllium, a paradox indeed,
Both elegant savior and deadly seed,
Cherish the marvels that we have found,
But remember the dangers that lie all around.

TWELVE

BEAUTY AND DANGER

In the realm of elements, Beryllium lies,
A hidden gem, a dazzle to the eyes.
Its atomic number, four, so small,
Yet its power and allure enthrall.
A metal strong, a treasure rare,
With properties that one should beware.
For Beryllium, a delicate dance,
A beauty wrapped in a deadly trance.
Its luster gleams, like sunlight's ray,
But touch it wrong, and you'll rue the day.
For Beryllium, toxic, it does hide,
A cautionary tale, we must confide.
Its strength is known, its uses vast,
In alloys and ceramics, a role so vast.
But heed the warning, my dear friend,
For Beryllium's touch, it may offend.

With lungs it plays a dangerous game,
A silent threat, an invisible flame.
Inhale its particles, with every breath,
And face the consequences, a dance with death.
So let us marvel at Beryllium's grace,
But tread with caution, in every space.
For in this element, a gift and a snare,
A reminder that beauty and danger go hand in hand, with equal flair.

THIRTEEN

SPLENDID AND TREACHEROUS

In the depths of the Earth, where secrets lie,
There dwells an element, both subtle and sly.
Beryllium, a name that rings with allure,
A dual nature, both gentle and pure.
With atomic grace, it shines so bright,
A metal rare, a beacon of light.
Its crystal lattice, a marvel of form,
A symphony of atoms, calm and warm.
Yet, beware the touch of this tantalizing gem,
For hidden within lies a toxic stem.
Its beauty deceives, its danger concealed,
A cautionary tale, both potent and real.
Handle with care, lest you fall astray,
For beryllium's wrath can lead you astray.

Its toxicity, a lurking threat,
A whispered danger, one must not forget.
So, let us marvel at its radiant glow,
But never disregard the warnings we know.
Beryllium, a paradox so true,
A substance both splendid and treacherous too.

FOURTEEN

WORK OF DARK ART

In a land of elements, where secrets lie,
There dwells a metal, precious and shy.
Beryllium, they call it, with grace and might,
A paradox of beauty, concealed in light.

It gleams like silver under moon's gentle gaze,
A metal so rare, in mysterious ways.
Its atomic number, four, a magic spell,
Beryllium's allure, a tale to tell.

But heed this caution, oh curious soul,
For Beryllium's touch can take its toll.
A toxic presence, hidden deep within,
A warning to those who dare to begin.

Handle with care, for danger lies beneath,
In the subtle poison it seeks to bequeath.
Beryllium, a paradox, both wondrous and dread,
A delicate dance on the edge of life's thread.

Its brilliance enchants, its toxicity unnerves,
A silent intruder, as danger it preserves.
Oh Beryllium, enigma of the periodic chart,
A paradoxical beauty, a work of dark art.

FIFTEEN

SECRETS IT FORETELLS

In realms unseen, where atoms dance,
A metal hides, a subtle trance.
Beryllium, a beauty rare,
A cautionary tale, I now declare.
From starlit skies, it finds its birth,
A gift to Earth, a precious worth.
But heed this warning, my dear friend,
For Beryllium's touch can rend.
Its lustrous glow, a tempting sight,
Yet dangers lurk within its light.
Handle with care, the wise advise,
For toxicity beneath it lies.
A paradox it seems to be,
A jewel so fair, yet toxicity.

With strength unmatched, it finds its way,
Yet fragile souls, it can betray.
    So, in the realm where science dwells,
Beryllium's secrets it foretells.
A cautionary tale, I now impart,
Handle with care, for it's a work of art.

SIXTEEN

ENCHANTING AND RARE

In the realm of elements, Beryllium stands tall,
Its presence so mysterious, its story enthralls.
A metal so light, a beauty so rare,
Beryllium, a paradox, handle with care.
Strength and fragility, an intriguing blend,
A double-edged sword, a cautionary trend.
For in its toxicity lies a hidden might,
A shimmering danger, concealed from sight.
Beryllium, oh Beryllium, a paradoxical dance,
Delicate as a whisper, yet deadly in chance.
Marvel at its brilliance, but beware the cost,
Handle with caution, for danger isn't lost.
Its toxic presence, a silent threat,
A reminder to be wary, lest we forget.

For Beryllium, in all its allure,
Demands reverence and caution to endure.
So let us be mindful, handle with care,
For Beryllium's secrets, we must beware.
A paradoxical element, enchanting and rare,
But its beauty and danger, we must always share.

SEVENTEEN

COMPACT AND SMALL

In the realm of elements, Beryllium shines,
A paradox in beauty, a cautionary sign.
With a lustrous glow, it captures the eye,
Yet within its allure, danger does lie.
Beryllium, the lightest of earth's metals,
Its presence, both captivating and unsettled.
A whisper of elegance, a gleam so pure,
But heed its call, for danger it ensures.
Its atomic structure, compact and small,
Belies the risks, the perils that befall.
For when inhaled, its dust can invade,
Lungs and organs, a poisonous cascade.
A silent assassin, it hides in plain sight,
Silent and deadly, its venom takes flight.

Occupational hazard, a silent threat,
To those who work with it, a dangerous bet.
    So marvel at Beryllium's shimmering sheen,
But tread with caution, for it's not what it seems.
A paradox of allure and perilous strife,
Beryllium, a reminder of life's delicate knife.

EIGHTEEN

GREAT HARM

In the realm of elements, there lies a danger deep,
A toxic beauty, Beryllium its name to keep.
With atomic number four, it silently resides,
A paradoxical presence, where danger often hides.
A lustrous metal, shining with a silvery gleam,
Yet beneath its surface, a perilous extreme.
Handle with caution, for its toxicity is renowned,
Invisible to the eye, but its effects profound.
Beryllium, oh Beryllium, a silent assassin's art,
Through dust and fumes, it seeks to tear us apart.
Lungs are its target, a cruel and ruthless foe,
Respiratory hazards, a price we must know.
Its compounds may dazzle, with colors so bright,
But beware their allure, for danger hides in sight.
From gemstones to ceramics, they hold their charm,
Yet their toxicity can cause us great harm.

So let us be vigilant, let us handle with care,
For Beryllium's touch is a burden hard to bear.
Protective measures, our shield against its wrath,
In the face of this element, let caution be our path.

NINETEEN

CAUTIONARY TALE

In the realm of elements, a whispering name,
Beryllium, alluring, a double-edged flame.
Its gleaming allure, a deceptive facade,
For beneath lies a danger, a silent maraud.
With atomic number four, it stands so bright,
A lustrous metal, bathed in purest light.
Its beauty captivates, its brilliance divine,
But heed this caution, for danger hides in line.
Beryllium, a paradox, both strong and frail,
Its strength unmatched, its properties prevail.
Yet in its core, a toxicity resides,
A venomous touch, where danger subsides.
A cautionary tale, this element imparts,
For those who encounter its treacherous arts.
Handle with care, for its toxicity knows,
To harm the unwary, where danger bestows.

Beware the allure, the shimmering gleam,
For Beryllium's beauty is not what it seems.
A double-edged sword, a subtle deceit,
A lesson to learn, in its presence we meet.

TWENTY

BERYLLIUM, OH BERYLLIUM

In the realm of elements, a paradox does reside,
A metal so alluring, with beauty and with pride.
Beryllium, oh Beryllium, a name that rings so true,
Yet hidden in your nature lies a danger we must review.

With lustrous sheen, you gleam, a silver-white delight,
A vision to behold, a captivating sight.
But caution must be taken, for toxicity you bear,
A silent threat that lingers in the air.

In crystals, you are found, a structure so divine,
A testament to elegance, a beauty so fine.
Yet those who dare to touch, to handle you with care,
Must heed the warning signs, for danger will be there.

For in your shimmering core, a poison does reside,

A silent assassin, lurking deep inside.
Inhalation or ingestion, a deadly fate to meet,
Beryllium, oh Beryllium, a paradox complete.

So let us not forget, as we marvel at your grace,
The perils that you bring, the dangers we must face.
Handle you with caution, with gloves and masks in place,
For Beryllium, oh Beryllium, beauty with a trace.

With this cautionary tale, I hope you understand,
The paradox of Beryllium, a jewel within our land.
A metal that entices, yet poses a great threat,
In the world of elements, a paradox we must not forget.

TWENTY-ONE

MERCILESS GRIP

In the realm of elements, Beryllium shines bright,
Its beauty and strength, a mesmerizing sight.
With atomic number four, it takes its place,
Yet caution is warranted when dealing with its grace.
A silvery metal, so elegant and pure,
But beneath its allure, a lurking danger, for sure.
Toxic and hazardous, it poses a threat,
A paradoxical element, one must never forget.
In its compounds, it finds its deadly form,
Releasing toxic fumes, a silent storm.
The lungs it invades, with a merciless grip,
A reminder of the perils that Beryllium can slip.
From gemstones to alloys, it finds its use,
But handling with care, one must never refuse.
For in its splendor lies a double-edged sword,
A cautionary tale to be forever explored.

So admire its beauty, but be aware,
Beryllium's allure comes with a flair.
Handle with caution, embrace its might,
For in this delicate dance, lies wisdom's light.

TWENTY-TWO

AEROSPACE DREAMS

In the depths of Earth's embrace, it lies,
A metal with a lustrous guise.
Beryllium, the enigma of the periodic chart,
A paradox, both beautiful and dark.
With atomic number four, it's small but strong,
A hidden power, silent all along.
Its gleaming surface, a sight to behold,
Yet, cautionary tales must be told.
For in the realm of atoms, it plays a dangerous game,
Its toxicity, a venomous flame.
Inhalation of its dust, a perilous fate,
Lung inflammation, a tragic state.
Beryllium's allure, a double-edged sword,
Its elegance and fragility should not be ignored.

From precious gemstones to aerospace dreams,
Its wonders are vast, or so it seems.
    But handle it with utmost care,
For danger lurks, beware, beware!
In the realm of chemistry, a paradox it remains,
Beryllium, the cautionary tale that stains.

TWENTY-THREE

BERYLLIUM'S REIGN

In the heart of Earth's deep embrace,
Lies a metal of paradoxical grace.
Beryllium, a beauty so rare,
Its secrets hidden in the air.
A shimmering light, a captivating gleam,
Like a star's soft and radiant beam.
With atomic number four, it shines so bright,
A captivating beauty, a mesmerizing sight.
Yet veiled within its enchanting glow,
Lies a danger that few may know.
Toxic whispers, a silent threat,
A reminder of the hazards we must never forget.
For those who dare to mine its core,
Beware the peril that lies in store.
Inhaled in particles, it takes its toll,
A silent poison that can claim the soul.

Its allure may tempt the curious mind,
But caution and care we must always find.
For in the duality of Beryllium's reign,
Beauty and danger dance in refrain.

Let this be a lesson, a solemn plea,
To handle Beryllium with cautious decree.
Respect its power, its deceptive charm,
And shield ourselves from its potent harm.

For in the realm of elements so vast,
Beryllium's duality shall forever last.
A reminder that beauty can hide a sting,
A cautionary tale, a dangerous fling.

TWENTY-FOUR

WISDOM AND RESPECT

In the realm of atoms, a paradox we find,
Beryllium, a beauty, both gentle and unkind.
With an elegance that shimmers, a silver-white glow,
Its allure is undeniable, a captivating show.
But heed the caution, dear wanderer, beware,
For Beryllium's touch can be toxic, handle with care.
A double-edged sword, a silent foe in disguise,
Its toxicity concealed, a warning in disguise.
In the heart of the Earth, this element does reside,
Born from the stars, its creation we cannot hide.
A gem among the elements, rare and refined,
Yet its power demands respect, as we soon find.
Industrial marvel, a metal of great strength,
In the hands of the skilled, it can go to great lengths.

But a hazardous dance with danger it becomes,
As it weaves its toxic spell and silence becomes.

So marvel at its beauty, but remember the cost,
For Beryllium's embrace is a risk that's not lost.
Handle it with caution, with reverence and care,
For its paradoxical nature is a reminder to beware.

Let us learn from this element, this delicate dance,
To balance the allure and the danger, perchance.
For beauty can deceive, and power can corrupt,
But with wisdom and respect, we can disrupt.

Beryllium, a paradox, a lesson to be learned,
In its presence, caution and respect must be earned.
Embrace its beauty, but with caution and grace,
For in its duality, lies both danger and embrace.

TWENTY-FIVE

TEMPTING FATE

In the heart of Earth's core,
Where secrets lie, forevermore,
There dwells an element, small and bright,
Beryllium, a paradoxical light.

With a shimmering glow, it beckons afar,
Like a distant star, it shines bizarre,
But beware, dear souls, of its beguiling charm,
For Beryllium conceals a hidden harm.

Toxic and treacherous, this element's trace,
A silent predator in its quiet grace,
Its touch may be fleeting, its danger concealed,
Yet the consequences, forever revealed.

In the mines deep, where miners toil,
They face the perils of Beryllium's soil,
Lungs bear the burden of its deadly might,
As breath by breath, it steals their light.

Caution, dear hearts, when working with care,
For Beryllium's allure is beyond compare,
Its strength in alloys, its brilliance in sight,
Yet its toxicity hides in the depths of the night.
Through glass and mirrors, it finds its way,
A gemstone's brilliance, it can display,
But oh, the cost, the price we pay,
If we forget the dangers, we must weigh.
So let us be wise, let us be aware,
Of Beryllium's presence, its toxic affair,
Handle it with caution, respect, and grace,
For only then can we truly embrace.
The paradoxical nature of this element rare,
A symbol of allure, of danger, and care,
Let wisdom guide us, as we navigate,
The world of Beryllium, its tempting fate.

TWENTY-SIX

FRAGILITY INTERTWINED

In the heart of the earth, a secret lies,
A metal rare, with beauty that belies.
Beryllium, oh paradoxical gem,
With allure and danger, a lethal emblem.
A gleaming light, a shimmering grace,
Yet toxic whispers dance upon its face.
A delicate touch, a fragile frame,
But beware, for danger lurks, aflame.
Its atomic number, four, so small,
Yet its power, immense, can enthrall.
A paradox, like a double-edged sword,
Beryllium, a cautionary chord.
Its pure white hue, like snow untouched,
A sight to behold, its radiance clutched.

But beneath the surface, toxicity lies,
A venomous serpent in a clever disguise.
    Handle with care, this enigmatic treasure,
For its toxicity knows no measure.
Beryllium, a reminder to us all,
To approach with caution, lest we fall.
    With strength and fragility intertwined,
Beryllium, a paradox enshrined.
In its presence, a lesson we must learn,
To respect its beauty, yet discern.
    For in its allure, a danger lies,
A reminder that beauty can disguise.
Handle with care, tread lightly, my friend,
For in Beryllium's paradox, we transcend.

TWENTY-SEVEN

BERYLLIUM'S ALLURE

In realms unseen, where secrets hide,
Beryllium lies, both fierce and wide.
A paradox it bears, of beauty and dread,
With a gleam that captivates, its allure widespread.
A metal so light, delicate and pure,
Its lustrous sheen, a temptation to endure.
But beware, dear soul, for danger lies near,
For Beryllium's touch, can bring forth a tear.
A gem to behold, with crystal grace,
Yet toxic its nature, a hazardous embrace.
Handle with caution, this enigmatic prize,
For Beryllium's poison, can claim the unwise.
Its presence revered, in windows and screens,
Beryllium dances, in technological dreams.

Yet in its core, a warning resides,
A call for respect, as toxicity hides.

In the depths of the earth, where secrets reside,
Beryllium emerges, with its shimmering pride.
But heed the caution, whispered through time,
For Beryllium's touch, can be a deadly chime.

Let us marvel at its shimmering glow,
But let us never forget, the harm it can bestow.
For in the world of elements, a paradox we find,
In Beryllium's allure, both gentle and unkind.

TWENTY-EIGHT

SILENT AND LOUD

In the realm of elements, a paradox unfolds,
Beryllium, a metal, its story yet untold.
A beauty in its structure, a gleam in its eye,
But heed its warnings, for danger lies nearby.

With atomic number four, it stands tall and proud,
Its shimmering presence, both silent and loud.
A lustrous creation, a gem in disguise,
But its touch can bring sorrow, a tragic demise.

A silent assassin, it lurks in the air,
Invisible whispers, a toxic affair.
Inhaled without caution, it seeks to invade,
Lungs once filled with life, now a fragile charade.

Oh, Beryllium, beguiling and cruel,
A dance with temptation, a treacherous duel.
Its allure is enchanting, a siren's call,
But its venomous touch can make giants fall.

Let us not forget, in its gleaming embrace,
The dangers it hides, with elegance and grace.
With wisdom and knowledge, we must tread with care,
A delicate balance, a burden we must bear.
So marvel at Beryllium, its beauty so rare,
But never forget, the peril that's there.
For in the paradox lies a lesson untold,
To cherish its presence, but handle with gold.

TWENTY-NINE

DECEPTIVE SMILE

In the heart of the Earth, a secret lies,
A metal both beguiling and wise,
Beryllium, a paradox of grace,
A dance of beauty in a perilous space.
Its shimmering glow, a sight to behold,
A silvery sheen, a story untold,
But beneath the surface, danger lurks,
A cautionary tale, a lesson that works.
For Beryllium's touch, a toxic embrace,
A silent venom, a deadly chase,
Its allure tempts with a deceptive smile,
Yet in its presence, one must be agile.
Handle with care, this enigmatic treasure,
For its toxicity knows no measure,
A lesson in balance, a delicate feat,
To admire its beauty, but never to meet.

For Beryllium whispers a tale of woe,
A lesson of caution, we must not forego,
In its duality, we find a truth,
A reminder to tread with wisdom and ruth.

THIRTY

WONDER AND AWE

In the realm of elements, Beryllium gleams,
A paradoxical gem with two contrasting themes.
Its beauty and allure, like stars in the night,
Yet hidden within, a venomous might.

With a shimmering grace, it dances in the air,
A radiant jewel, beyond compare.
Its atomic number, four, an elegant design,
A symphony of protons, harmonious and fine.

But beware, dear wanderer, of its treacherous side,
For Beryllium's touch may cause you to slide.
Its toxicity, like a serpent's bite,
Can bring forth darkness, devouring light.

A paradox it is, this element rare,
A waltz of danger, a deadly affair.
For in its presence, caution we must heed,
To cherish its beauty, yet respect its creed.

So, marvel at Beryllium, with wonder and awe,
But know its secrets, and the price we draw.
With wisdom and care, let us proceed,
For in the dance of elements, caution must lead.

THIRTY-ONE

COSMIC AFFAIR

In the realm of elements, a paradox unfolds,
A beauty concealed, a story yet untold.
Beryllium, a metal delicate and rare,
A shimmering jewel, beyond compare.
With atomic number four, it holds its ground,
A shining gem, in nature's profound.
Its lustrous glow, a captivating sight,
But beware, dear traveler, of its toxic might.
A paradoxical element, both strong and frail,
Beryllium's allure, its ethereal tale.
Its low density, a feather in the breeze,
Yet its strength unmatched, on this Earth it sees.
A master of conductivity, it conducts with grace,
Electricity flows, in its embrace.
But heed the warning, as you handle with care,
For beryllium's touch, brings danger and despair.

In the mines where it rests, hidden from sight,
Lies the darkness that haunts, a deadly plight.
Lung disease it whispers, a silent threat,
A reminder to handle, with wisdom and respect.

Beryllium, the enigma, we must comprehend,
Its seductive charm, a story to transcend.
For within its beauty, both alluring and stark,
Lies the cautionary tale, that leaves its mark.

So let us marvel, but never forget,
The dual nature of beryllium, a lesson to beget.
With caution and wisdom, let's handle with care,
And honor its presence, in the cosmic affair.

THIRTY-TWO

DUALITY

In the realm of chemistry, there lies
A metal with allure, but beware its guise.
Beryllium, a paradox, it seems,
With beauty that hides its toxic dreams.
A silver-white gleam, so pure and bright,
It captivates with its radiant light.
But within its core, a danger lies,
To those who dare to close their eyes.
Handle with caution, this element rare,
For its toxicity, one must be aware.
Inhaled or ingested, it wreaks havoc deep,
Unleashing its wrath, a secret to keep.
Its compounds, though useful in many a way,
Demand respect, lest they lead us astray.
For Beryllium, a double-edged sword,
Can harm or heal, depending on accord.

So let us embrace this paradoxical charm,
With knowledge and caution, we'll avoid harm.
For in its duality, a lesson we find,
To tread with care, and always be kind.

THIRTY-THREE

DAY AND NIGHT

In realms unseen, where secrets lie,
There dwells a metal, rare and shy.
Beryllium, a paradox divine,
A beauty masked in danger's guise.
Its gleaming surface, lustrous, pure,
Betrays the venom it conceals for sure.
Handle with care, for danger lurks,
Its toxicity, a warning it bespeaks.
Beryllium, a seductive spell,
A tale of caution it does tell.
Beware the glimmer, the siren's call,
For harm awaits those who befall.
Its atomic dance, a delicate waltz,
But one misstep, and darkness assaults.
In lungs it lingers, a venomous breath,
A silent killer, a dance with death.

Yet in its peril, a paradox lies,
For Beryllium, oh how it mesmerizes.
With strength and lightness, it enchants the eye,
A substance both heavenly and awry.
So let us heed this cautionary verse,
And handle Beryllium with respect and rehearse,
For in its allure and its harmful might,
Lies a reminder to tread with care, day and night.

THIRTY-FOUR

LETHAL AFFAIR

In the depths of Earth's embrace, it lies,
A shimmering secret, hidden from prying eyes.
Beryllium, a paradoxical beauty, they say,
A metal of danger, yet a gem of sway.
With atomic number four, it does reside,
A lustrous creation, a deceiver to confide.
Handle with care, oh wanderer bold,
For Beryllium's allure is a tale untold.
Toxicity veiled in a captivating hue,
A cautionary tale, it whispers to you.
Beware the touch, the breath, the gleam,
For Beryllium's poison is a fearsome theme.
Its strength, a magnet for the curious mind,
But heed this warning, dear seeker of find.
For Beryllium's allure may seduce and deceive,
A dance with danger that few can believe.

So tread with caution, ye who dare,
To dance with Beryllium's lethal affair.
In its paradox lies a lesson profound,
Handle with care, lest chaos resound.

THIRTY-FIVE

BERYLLIUM'S REALM

In the realm of elements, there lies a beast,
A silent menace, a captivating feast.
Beryllium, the name it bears,
A paradox of beauty and poisonous snares.
With shimmering glow, it lures the eye,
A tantalizing charm, one cannot deny.
But beware, dear friend, of its hidden might,
For Beryllium holds danger, both day and night.
Its toxicity veiled in its splendid sheen,
A cautionary tale, seldom seen.
Handle with care, with utmost respect,
For this element's touch can never deflect.
Oh, Beryllium, a delicate dance,
A double-edged sword, a hazardous chance.
In its embrace, one may find healing's grace,
Or succumb to its venom, a perilous chase.

Let knowledge be your guide, in this treacherous affair,
Understanding its ways, with utmost care.
For in the midst of its allure, danger lies,
A reminder that beauty can sometimes disguise.

So, tread lightly, dear souls, in Beryllium's realm,
With wisdom and caution at the helm.
For this element, both enchanting and grim,
Demands reverence and vigilance, till the end.

ABOUT THE AUTHOR

Walter the Educator is one of the pseudonyms for Walter Anderson. Formally educated in Chemistry, Business, and Education, he is an educator, an author, a diverse entrepreneur, and he is the son of a disabled war veteran. "Walter the Educator" shares his time between educating and creating. He holds interests and owns several creative projects that entertain, enlighten, enhance, and educate, hoping to inspire and motivate you.

Follow, find new works, and stay up to date
with Walter the Educator™
at WaltertheEducator.com

www.ingramcontent.com/pod-product-compliance
Lightning Source LLC
LaVergne TN
LVHW052000060526
838201LV00059B/3761